Department Store Math
for Beginners

REM 161D

A TEACHING RESOURCE FROM...

REMEDIA PUBLICATIONS

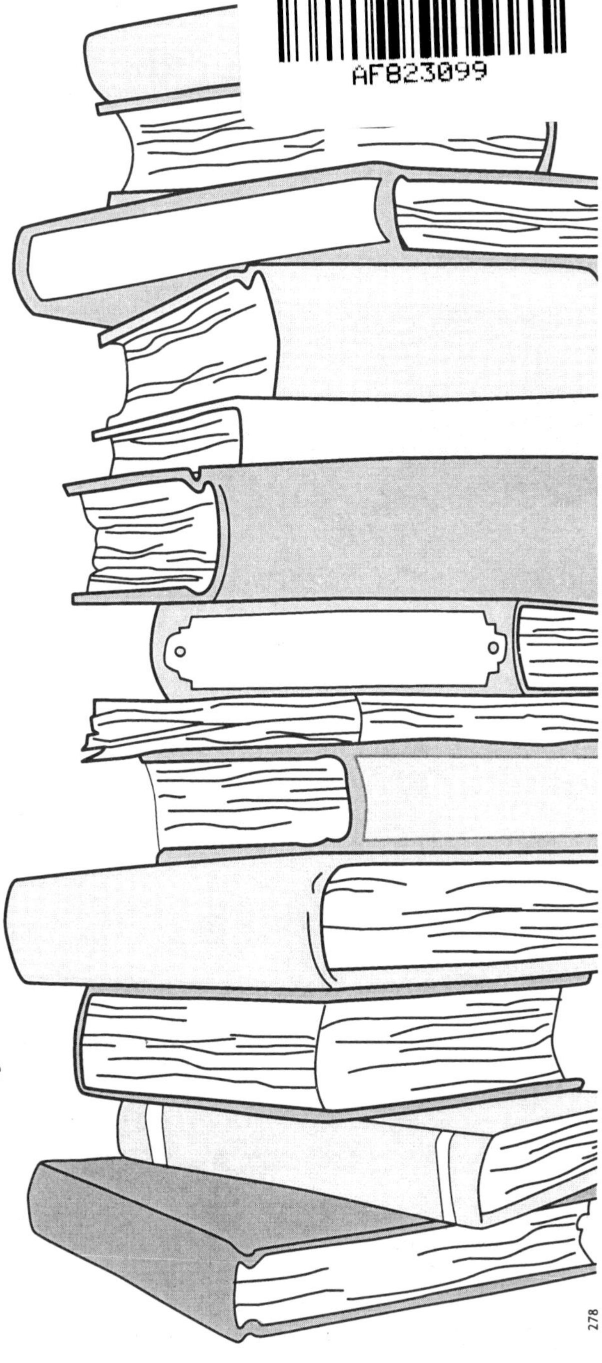

To find Remedia products in a store near you, visit:
http://www.rempub.com/stores

REMEDIA PUBLICATIONS, INC.
15887 N. 76TH STREET • SUITE 120 • SCOTTSDALE, AZ • 85260

DEPARTMENT STORE MATH

FOR BEGINNERS

To the teacher:

This book has been created to help students build basic math skills through the use of real-life situations (looking at a list of department store items and prices and figuring costs involved when shopping at the department store).

This approach has proven successful in helping to motivate the sometimes reluctant learner. Most students show a natural interest in materials featuring topics familiar to them as part of their everyday lives.

Not only do students who use *Department Store Math for Beginners* have an opportunity to improve basic math skills, but they are provided with practice in skimming for information.

-- Suggestions --

1. Laminate the department store price list.

2. Make additional price lists available so that several students can use the pages at the same time (price lists are available for purchase in packages of 6).

3. Set up a *Department Store Math* Learning Center.

4. Use play money to help students solve math problems.

5. Have students create their own department store math problems.

Name______________________________

Use the price list. Find the price of each item.

1.  $______________

2.  $______________

3.  $______________

4.  $______________

5.  $______________

6.  $______________

7.  $______________

8.  $______________

9.  $______________

10.  $______________

11.  $______________

12.  $______________

13. 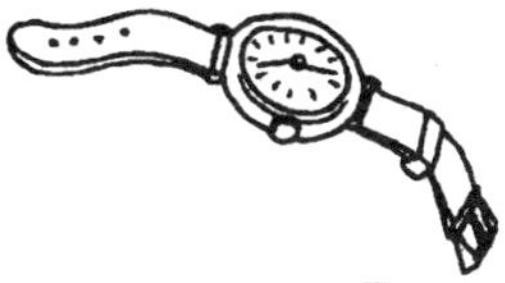 $______________

14. 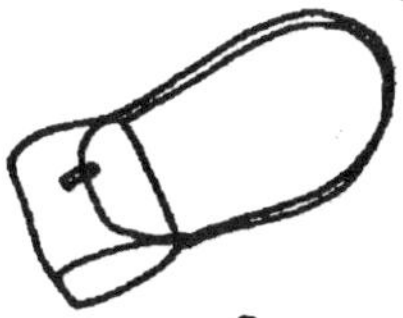 $______________

15.  $______________

16.  $______________

17.  $______________

18.  $______________

19.  $______________

20. 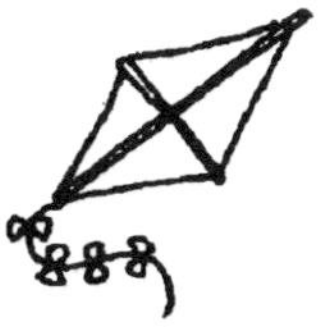 $______________

21.  $______________

22.  $______________

Name________________________________

Use the price list. Find the price of each item.

1. Wallet $__________
2. Backpack $__________
3. Skirt $__________
4. Camera $__________
5. Coloring book $__________
6. Swimsuit (girl's) $__________
7. Jump rope $__________
8. T-shirt $__________
9. In-line skates $__________
10. Dress $__________
11. Mittens $__________
12. Sneakers $__________
13. Bracelet $__________
14. Doll $__________
15. Jeans $__________
16. Fishing pole $__________
17. Blouse $__________
18. Sunglasses $__________
19. Frisbee $__________
20. Sweater $__________
21. Basketball $__________
22. Skateboard $__________

Name____________________________________

ADDITION - 2 items

Use the price list. Find the total cost.

1.

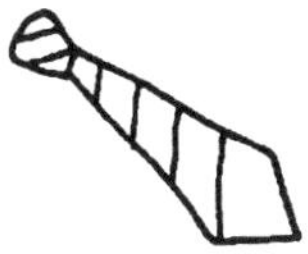

$______________

$______________

Total $______________

2.

$______________

$______________

Total $______________

3.

$______________

$______________

Total $______________

4.

$______________

$______________

Total $______________

5.

$______________

$______________

Total $______________

6.

$______________

$______________

Total $______________

7.

$______________

$______________

Total $______________

8.

$______________

$______________

Total $______________

Name________________________________

ADDITION - 2 items

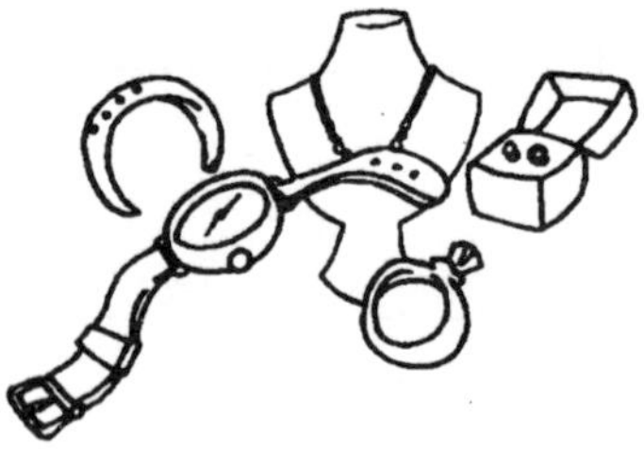

Use the price list. Find the total cost.

1.  

$______________

2.  

$______________

3.  

$______________

4.  

$______________

5.  

$______________

6.  

$______________

7.  

$______________

8. 

$______________

Name________________________________

ADDITION - 2 items

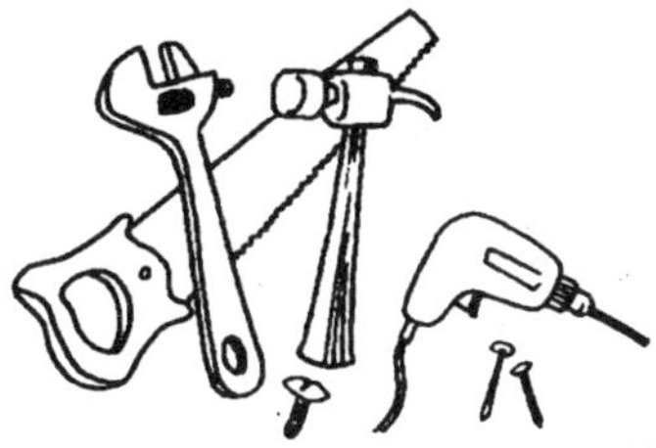

Use the price list. Find the total cost.

1. 

$______________

2. 

$______________

3. 

$______________

4. 

$______________

5. 

$______________

6. 

$______________

7. 

$______________

8. 

$______________

9. 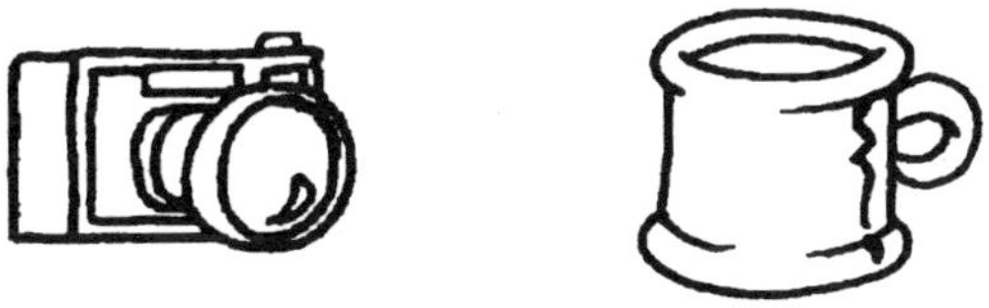

$______________

10. 

$______________

Name____________________________

ADDITION - 2 items

Use the price list. Solve each problem.

---

1. One  and one  cost $__________

2. One  and one  cost $__________

3. One  and one  cost $__________

4. One  and one  cost $__________

5. One  and one  cost $__________

6. One  and one 

cost $__________

7. One  and one 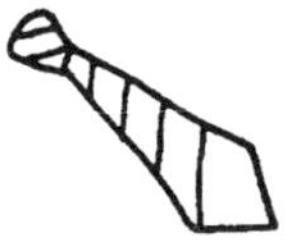 cost $__________

8. One  and one  cost $__________

Name________________________________

ADDITION - 2 items

Use the price list. Find the total cost.

1. 

$____________

$____________

Total $____________

2. 

$____________

$____________

Total $____________

3. 

$____________

$____________

Total $____________

4. 

$____________

$____________

Total $____________

5. 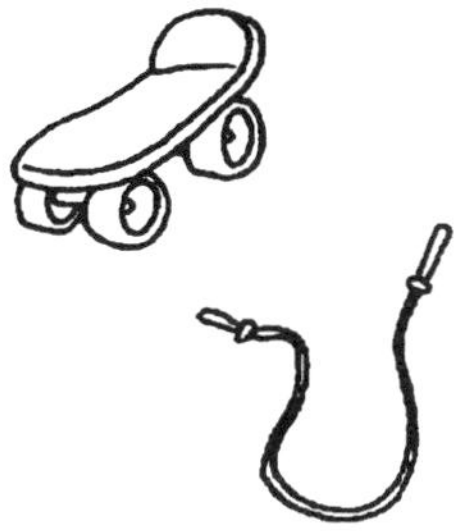

$____________

$____________

Total $____________

6. 

$____________

$____________

Total $____________

7. 

$____________

$____________

Total $____________

8. 

$____________

$____________

Total $____________

Name____

ADDITION - 2 items

Use the price list. Find the total cost.

1.  

$______________

2.  

$______________

3.  

$______________

4.  

$______________

5.  

$______________

6.  

$______________

7. 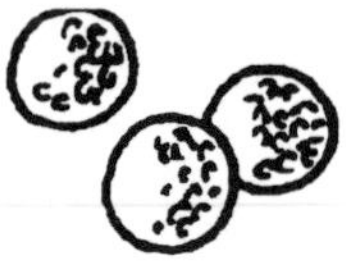 

$______________

8.  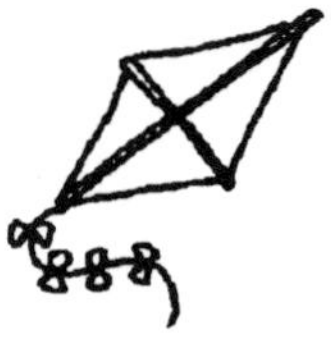

$______________

9. 

$______________

10.  

$______________

Name ___

ADDITION - 2 items

Use the price list. Find the total cost.

1.  $ ___

 $ ___

Total $ ___

2.  $ ___

 $ ___

Total $ ___

3.  $ ___

 $ ___

Total $ ___

4.  $ ___

 $ ___

Total $ ___

5.  $ ___

 $ ___

Total $ ___

6.  $ ___

 $ ___

Total $ ___

7.  $ ___

 $ ___

Total $ ___

8.  $ ___

 $ ___

Total $ ___

Name________________________________

ADDITION - 2 items

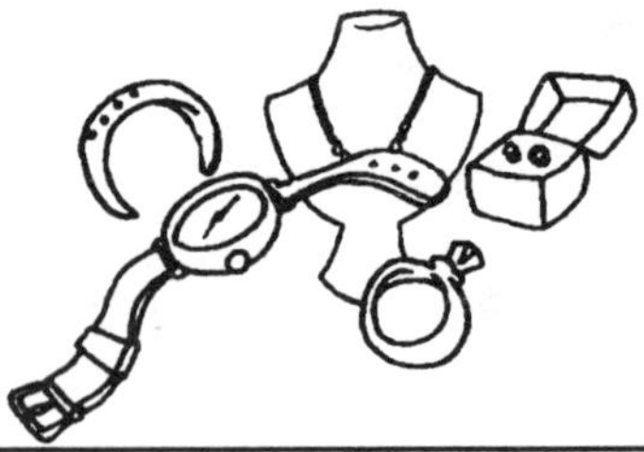

Use the price list. Solve each problem.

---

1. One  and one  cost $____________

2. One  and one  cost $____________

3. One  and one  cost $____________

4. One  and one  cost $____________

5. One  and one  cost $____________

6. One  and one  cost $____________

7. One  and one  cost $____________

8. One  and one  cost $____________

Name________________________________

ADDITION - 3 items

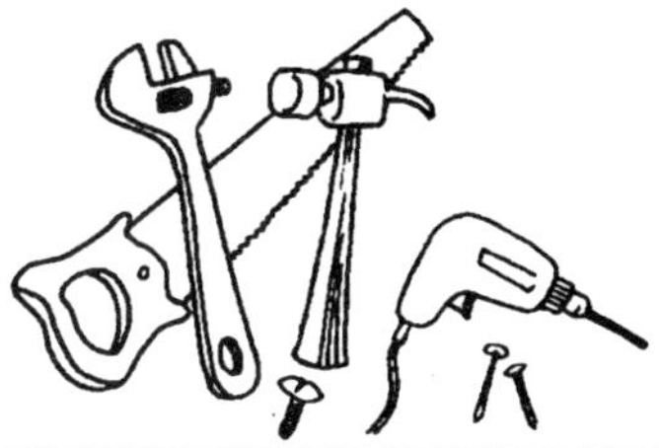

Use the price list. Find the total cost.

1.

 $____________

 $____________

 $____________

Total $____________

4.

 $____________

 $____________

 $____________

Total $____________

2.

 $____________

 $____________

 $____________

Total $____________

5.

 $____________

 $____________

$____________

Total $____________

3.

 $____________

 $____________

 $____________

Total $____________

6.

 $____________

 $____________

 $____________

Total $____________

Name____

ADDITION - 3 items

Use the price list. Find the total cost.

1. 

$________________

2. 

$________________

3. 

$________________

4. 

$________________

5. 

$________________

6. 

$________________

7. 

$________________

8. 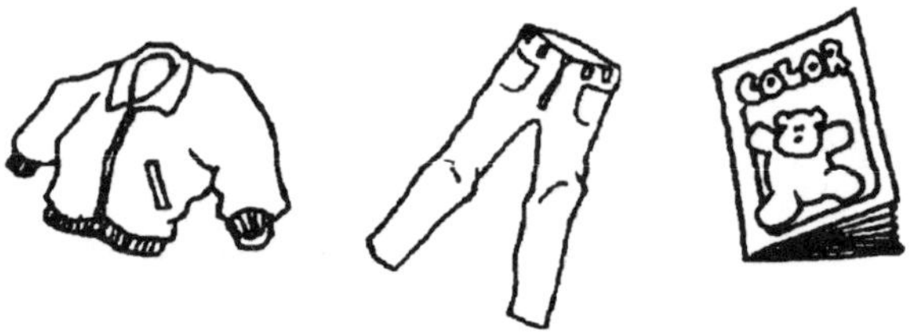

$________________

Name____

ADDITION - 3 items

Use the price list. Solve each problem.

---

1. Sally bought  .

   How much in all? $____________

2. What is the total cost of  ? $____________

3. Mrs. Baker bought  and  .

   How much did she spend? $____________

4. How much money do you need to buy

    ? $____________

5. Nick bought  and  .

   How much did he spend? $____________

6. How much for 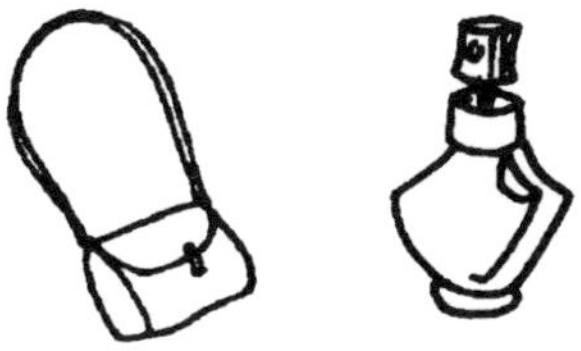 and  ? $____________

Name________________________________

ADDITION - 3 items

Use the price list. Find the total cost.

1. baseball cap $________ jump rope $________ purse $________ Total $________	4. basketball $________ sunglasses $________ coffee mug $________ Total $________
2. beach towel $________ tennis racquet $________ baseball glove $________ Total $________	5. in-line skates $________ dress $________ bracelet $________ Total $________
3. sneakers $________ swimsuit (boy's) $________ coloring book $________ Total $________	6. necklace $________ socks $________ crayons $________ Total $________

Name____

ADDITION - 3 items

Use the price list. Find the total cost.

1.

 $______________

 $______________

 $______________

Total $______________

4.

 $______________

 $______________

 $______________

Total $______________

2.

 $______________

 $______________

 $______________

Total $______________

5.

 $______________

 $______________

 $______________

Total $______________

3.

 $______________

 $______________

 $______________

Total $______________

6.

 $______________

 $______________

 $______________

Total $______________

Name______________________________

ADDITION - 3 items

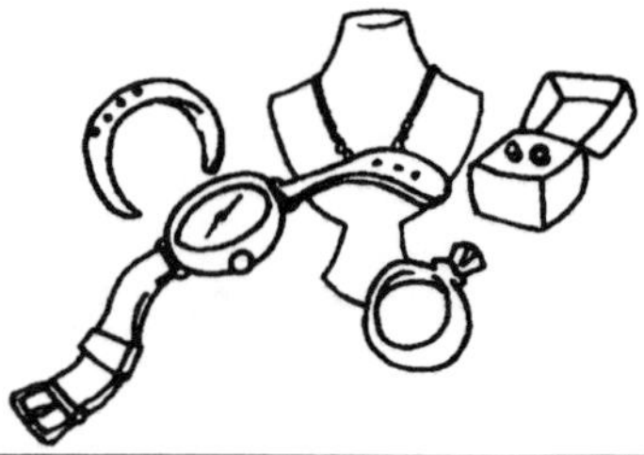

Use the price list. Find the total cost.

1. 

$______________

2. 

$______________

3. 

$______________

4. 

$______________

5. 

$______________

6. 

$______________

7. 

$______________

8. 

$______________

Name________________________________

ADDITION - 3 items

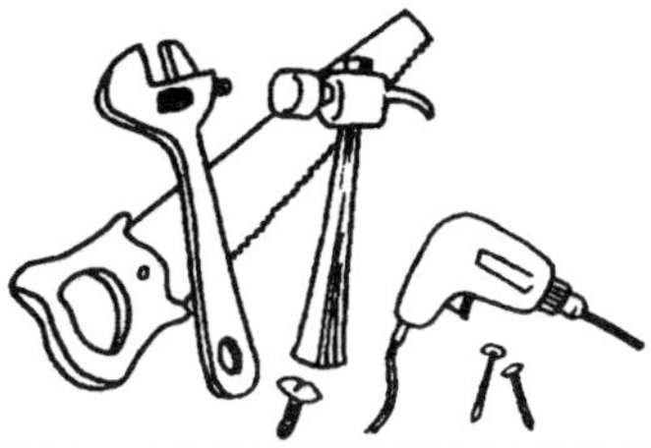

Use the price list. Solve each problem.

1. Beth bought a necklace, earrings, and a dress. How much did she spend? $______________

2. Josh bought his dad a necktie, a book, and a pair of socks for his birthday. What was the total cost? $______________

3. How much would you need to buy a camera, a kite, and a pair of sneakers? $______________

4. Sammy asked his mom for crayons, a coloring book, and a kite. What is the total cost? $______________

5. Mrs. Young bought a girl's swimsuit, a pair of sandals, and a beach towel. How much? $______________

6. What is the total cost for a baseball glove, a T-shirt, and a baseball cap? $______________

Name____

ADDITION - 3 items

Use the price list. Solve each problem.

---

1.   and  cost $______________

2.   and  cost $______________

3.   and  cost $______________

4.   and  cost $______________

5.   and  cost $______________

6.   and  cost $______________

7.   and  cost $______________

8.   and  cost $______________

Name________________________________

ADDITION - 2 & 3 items

Use the price list. Find the total cost.

1. One  and one  cost $______________

2. One  and two  cost $______________

3. One  and one  cost $______________

4. One  and one  cost $______________

5. One  and one  cost $______________

6. One  and two  cost $______________

7. One  and two  cost $______________

8. One  and two  cost $______________

Name_______________________________

SUBTRACTION

Use the price list. Solve each problem.

HAD	BOUGHT	HAD LEFT
1. $ 9.50		
2. $15.00		
3. $12.59		
4. $17.69		
5. $ 8.64		
6. $ 9.99		
7. $ 2.48		
8. $13.59		

Name________________________________

SUBTRACTION

Use the price list. Solve each problem.

1. Had $9.89

Bought

How much left? $____________

2. Had $9.89

Bought

How much left? $____________

3. Had $4.79

Bought

How much left? $____________

4. Had $5.75

Bought

How much left? $____________

5. Had $9.97

Bought

How much left? $____________

6. Had $9.89

Bought

How much left? $____________

7. Had $8.69

Bought

How much left? $____________

8. Had $9.88

Bought

How much left? $____________

Name______________________________

SUBTRACTION

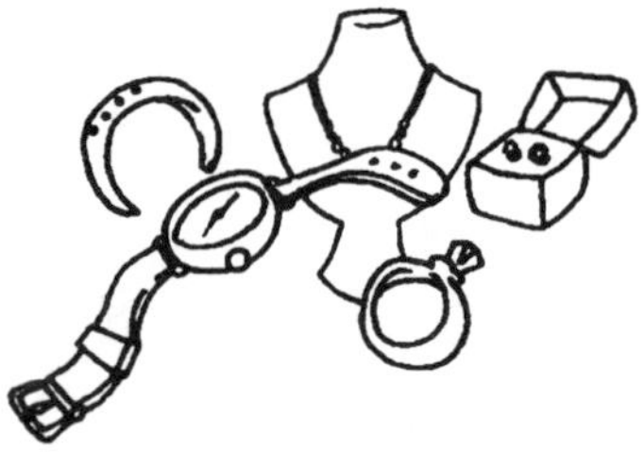

Use the price list. Solve each problem.

YOU WANT	YOU HAVE	HOW MUCH MORE DO YOU NEED?
1. 	$12.00	
2. 	$ 1.45	
3. 	$ 2.19	
4. 	$10.00	
5. 	$ 3.10	
6. 	$ 2.00	
7. 	$ 4.30	
8. 	$ 6.33	

Name____

SUBTRACTION

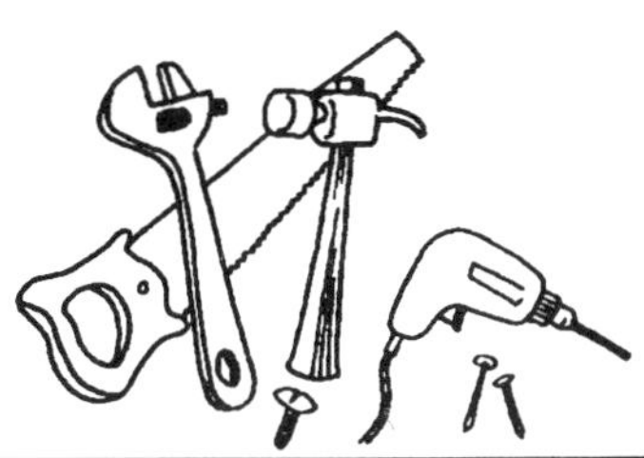

Use the price list. Solve each problem.

1. Tom bought

Jerry bought

How much more did Tom spend? $______________

2. Lucy bought

Ethel bought

How much more did Lucy spend? $______________

3. Lisa bought

Bart bought

How much more did Lisa spend? $______________

4. Tony bought

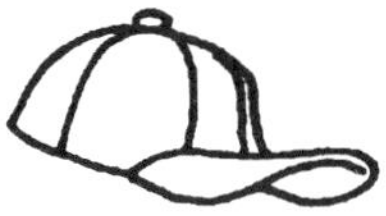

Rocky bought

How much more did Tony spend? $______________

5. Mickey bought

Donald bought

How much more did Mickey spend? $______________

6. Charlie bought

Linus bought

How much more did Charlie spend? $______________

Name____________________________________

SUBTRACTION

Use the price list. Solve each problem.

WHICH COSTS MORE? (circle your answer)				HOW MUCH MORE?
1.		or		$____________
2.		or		$____________
3.	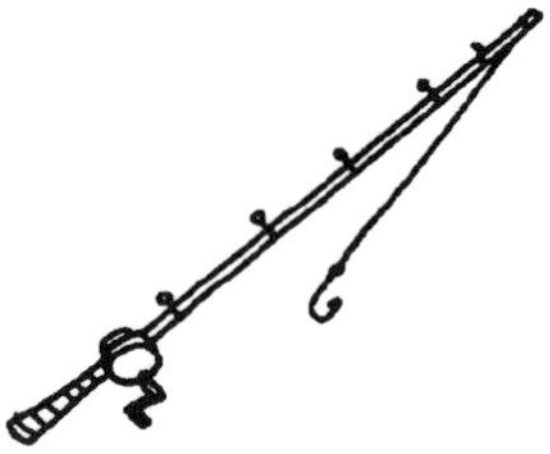	or		$____________
4.		or		$____________
5.		or		$____________
6.		or		$____________

Name______________________________

SUBTRACTION

Use the price list. Solve each problem.

1.

$______________ $______________

How much more for?

$______________

4.

 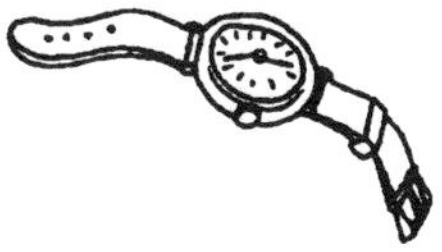

$______________ $______________

How much more for?

$______________

2.

$______________ $______________

How much more for?

$______________

5.

$______________ $______________

How much more for?

$______________

3.

 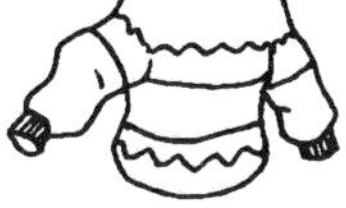

$______________ $______________

How much more for?

$______________

6.

$______________ $______________

How much more for?

$______________

Name____

SUBTRACTION

Use the price list. Solve each problem.

YOU WANT	YOU HAVE	HOW MUCH MORE DO YOU NEED?
1. 	$ 2.75	
2. 	$ 4.80	
3. 	$ 1.85	
4. 	$ 8.36	
5. 	$ 5.48	
6. 	$10.50	
7. 	$ 8.74	
8. 	$ 2.90	

Name________________________________

SUBTRACTION

Use the price list. Solve each problem.

---

1. Margo had $8.45. She bought a  .

   How much money left? $______________

2. Brad had $12.35. He bought a  .

   How much money left? $______________

3. Nick had $4.55. He bought  .

   How much money left? $______________

4. Sally had $7.22. She bought  .

   How much money left? $______________

5. Kelly had $4.58. She bought  .

   How much money left? $______________

6. Mike had $6.35. He bought a  .

   How much money left? $______________

Name______________________________

SUBTRACTION

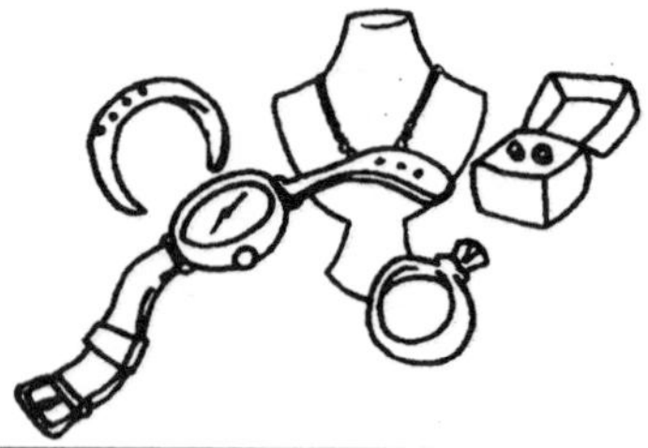

Use the price list. Solve each problem.

---

1. Nan bought a necklace. Sue bought a bracelet.

   How much more did Nan spend? $______________

2. Bud bought a camera. Beth bought a radio.

   How much more did the radio cost? $______________

3. Coach West had $15.00. He bought a tennis racquet.

   How much money does he have left? $______________

4. Tina wants a new purse. She has $8.75.

   How much more money does she need? $______________

5. Mrs. Green bought perfume. Mr. Green bought a necktie.

   How much more did Mr. Green spend? $______________

6. How much more does a girl's swimsuit cost

   than a boy's swimsuit? $______________

Name________________________________

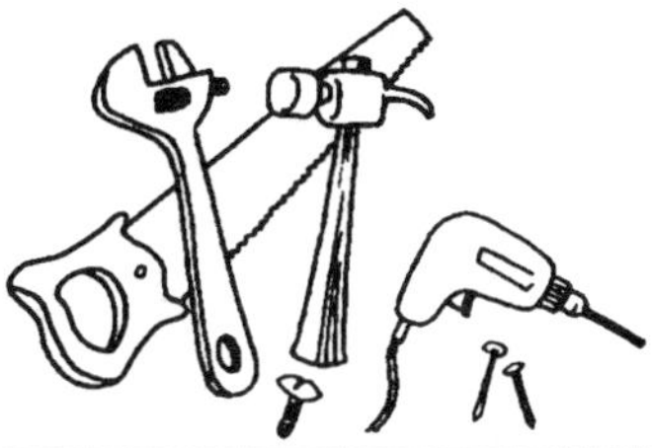

Use the price list. Solve each problem.

---

1. Joan bought a  and a  .

   a) How much did she spend? $______________

   b) How much change from $7.00? $______________

2. Rick bought a  and  .

   a) How much did he spend? $______________

   b) How much change from $15.00? $______________

3. Mrs. James bought a  and  .

   a) How much did she spend? $______________

   b) How much change from $12.00? $______________

4. Jay bought a  and a  .

   a) How much did he spend? $______________

   b) How much change from $40.00? $______________

Name____________________________________

ADDITION & SUBTRACTION

Use the price list. Solve each problem.

1. Bought  

Had $15.75

How much left? $______________

2. Bought  

Had $24.99

How much left? $______________

3. Bought  

Had $35.00

How much left? $______________

4. Bought  

Had $12.85

How much left? $______________

5. Bought  

Had $ 6.45

How much left? $______________

6. Bought  

Had $18.95

How much left? $______________

7. Bought  

Had $21.30

How much left? $______________

8. Bought  

Had $13.48

How much left? $______________

Name________________________________

ADDITION & SUBTRACTION - Zero

Use the price list. Solve each problem.

HAD	BOUGHT	HAD LEFT
1. $20.00	 	
2. $10.00	 	
3. $15.00	 	
4. $ 4.00	 	
5. $ 7.00	 	
6. $ 8.00	 	
7. $ 7.00	 	
8. $ 6.00	 	

Name____________________________________

ADDITION & SUBTRACTION

Use the price list. Solve each problem.

MONEY TO START WITH		BOUGHT		MONEY LEFT
1.	$20.00			
2.	$10.00			
3.	$15.00	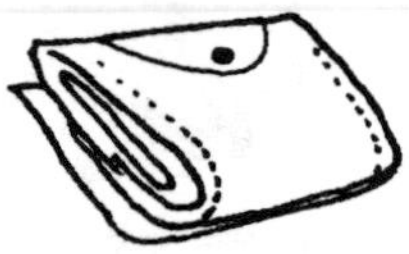		
4.	$14.00			
5.	$40.00			
6.	$18.00			
7.	$ 7.00			
8.	$16.00			

Name________________________________

Use the price list. Solve each problem.

---

1. Sue had $14.98. She bought a purse and sunglasses.

   How much money is left? $________________

2. Bud had $20.00. He wants to buy a skateboard and a watch.

   How much more money does he need? $________________

3. Jan has $7.50. She wants to buy a bracelet and earrings.

   How much money will be left? $________________

4. Mrs. Burns used $20.00 to pay for a purse and a camera.

   What was her change? $________________

5. Mr. Jones bought a tennis racquet and a box of golf balls.

   He gave the clerk $15.75.

   How much change did he get? $________________

6. Missy had $5.80. She bought a jump rope and a kite.

   How much money is left? $________________

Name________________________________

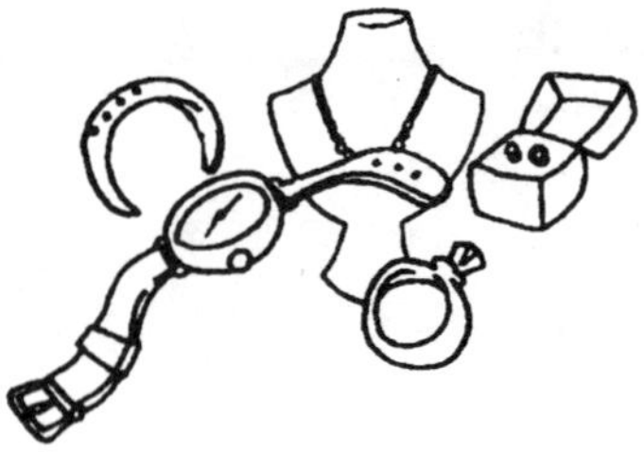

MULTIPLICATION

Use the price list. Solve each problem.

---

1. Two  cost $____________

2. Three 

cost $____________

3. Two  cost $____________

4. Three  cost $____________

5. Two  cost $____________

6. Three  cost $____________

7. Three  cost $____________

8. Two  cost $____________

9. Three  cost $____________

10. Two  cost $____________

Name_______________________________

MULTIPLICATION

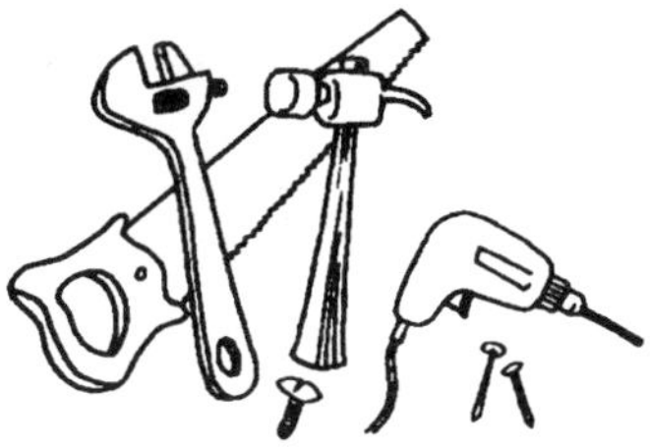

Use the price list. Solve each problem.

---

1. Two  cost $__________

2. Three  cost $__________

3. Four  cost $__________

4. Five  cost $__________

5. Four  cost $__________

6. Three  cost $__________

7. Four  cost $__________

8. Five  cost $__________

9. Two  cost $__________

10. Three  cost $__________

Name________________________________

MULTIPLICATION

Use the price list. Solve each problem.

---

1. Four  cost $____________

2. Six  cost $____________

3. Two  cost $____________

4. Five  cost $____________

5. Three  cost $____________

6. Two  cost $____________

7. Six  cost $____________

8. Five 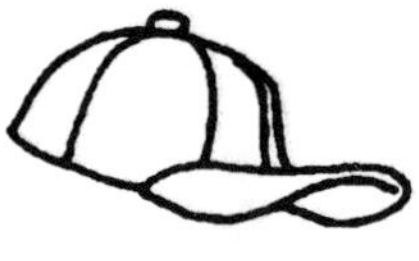 cost $____________

9. Two  cost $____________

10. Four  cost $____________

Name____

MULTIPLICATION

Use the price list. Solve each problem.

---

1. Coach West bought 3 baseball gloves.

   How much did he spend? $______________

2. Mrs. Mills bought 4 boxes of crayons.

   How much did she spend? $______________

3. What is the cost of 7 coloring books? $______________

4. Nancy bought 2 blouses.

   How much did she spend? $______________

5. How much for 5 beach towels? $______________

6. Mr. Gray bought 4 pairs of mittens.

   How much did he spend? $______________

7. Find the cost of 5 stuffed bears. $______________

8. Liz bought 6 pairs of socks.

   How much did she spend? $______________

Name________________________________

MULTIPLICATION

Use the price list. Find the total cost.

1. 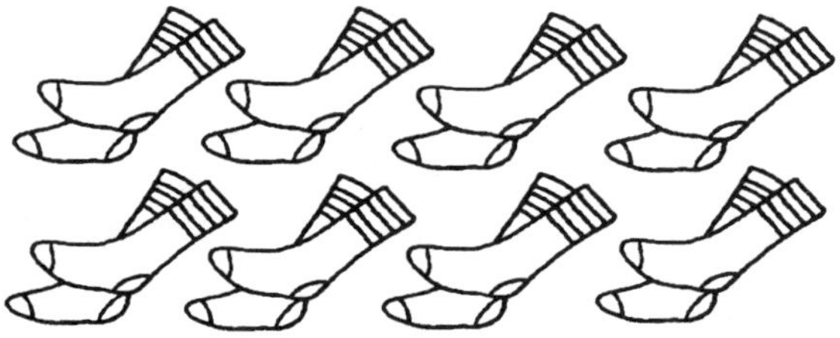

$______________

2. 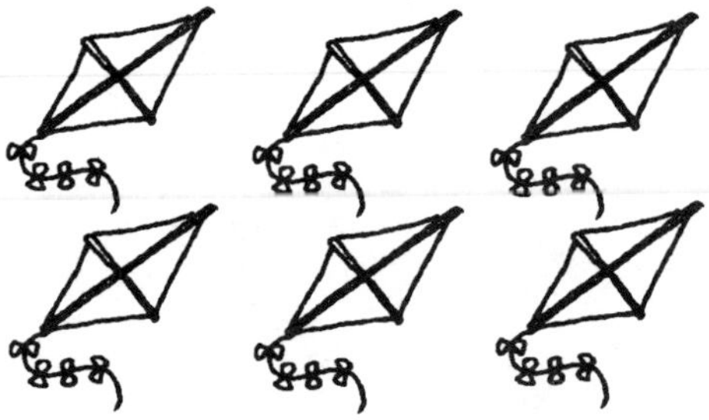

$______________

3. 

$______________

4. 

$______________

5. 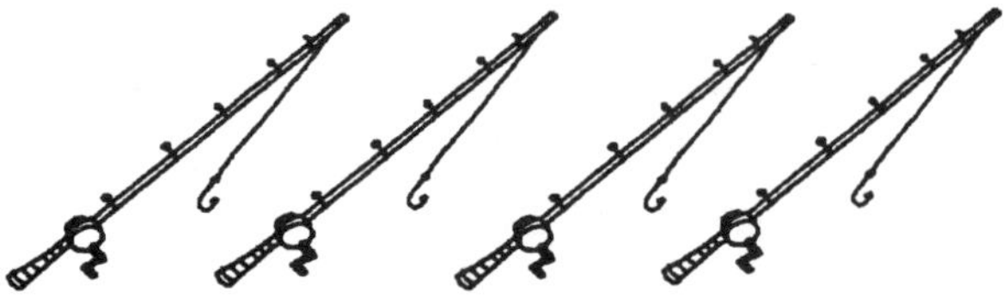

$______________

6. 

$______________

7. 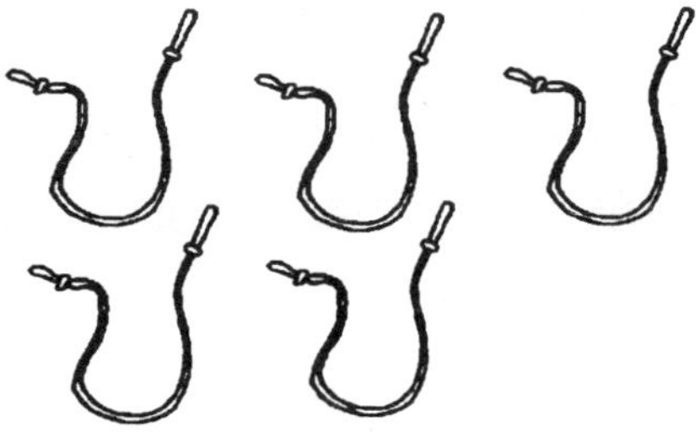

$______________

8. 

$______________

Name____

ADDITION, SUBTRACTION, MULTIPLICATION

Use the price list. Solve each problem.

1.

Bill bought

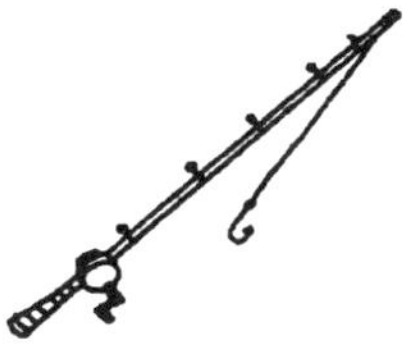

Ned bought

How much more did Ned spend? $______________

4.

$______________

$______________

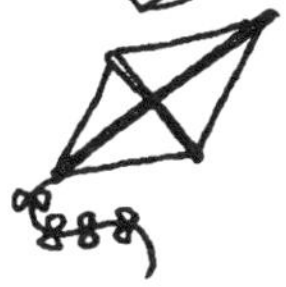

$______________

Total $______________

2.

Total $______________

5.

Sara bought

Lisa bought

How much more did Sara spend? $______________

3.

$______________

$______________

$______________

Total $______________

6.

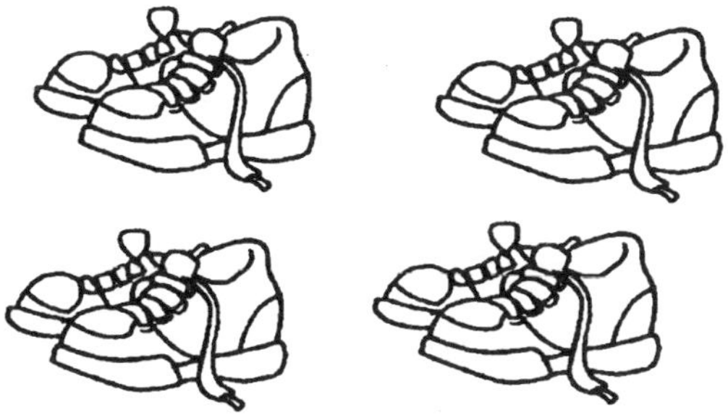

Total $______________

Name________________________________

ADDITION, SUBTRACTION,
MULTIPLICATION

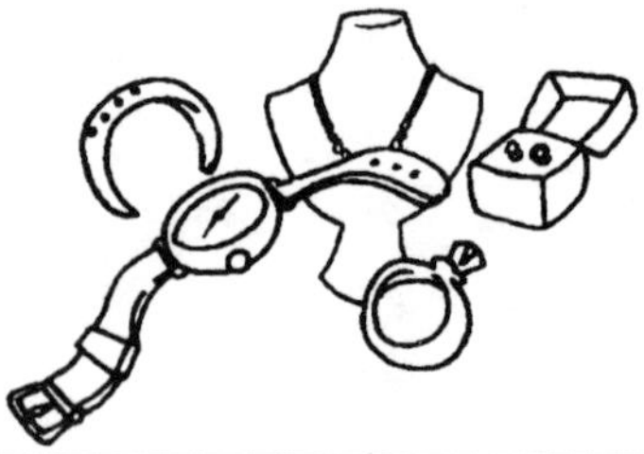

Use the price list. Solve each problem.

---

1. What is the total cost of    ? $____________

2. Jeff had $11.47.

   He bought  . How much left? $____________

3. Jan bought 3 . What was the total cost? $____________

4. Find the total cost of 2  and a . $____________

5. How much more are  than a  ? $____________

6. Steffie wants to buy  and  .

   She has $5.59.

   How much more money does she need? $____________

Name________________________________

ADDITION, SUBTRACTION,
MULTIPLICATION

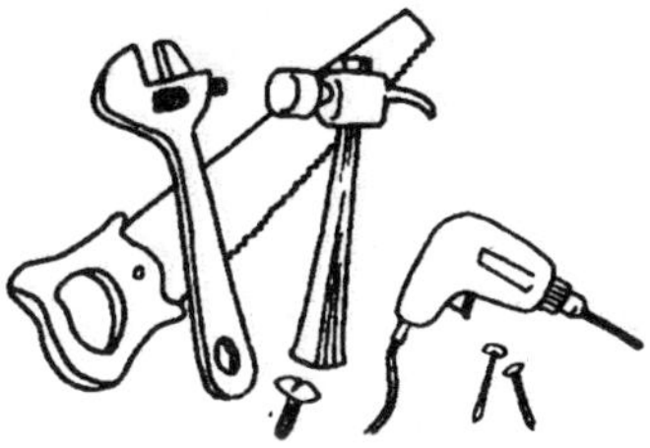

Use the price list. Solve each problem.

---

1. Find the cost of   and  . $______________

2. How much more for  than  ? $______________

3. Dan has $11.48 . He wants to buy  and  . How much more does he need? $______________

4. Find the total cost of 2  and  . $______________

5. Mrs. Fry had $20.00. She bought 2  and a  . How much money is left? $______________

6. Nicky bought   and  . How much money did she spend? $______________

Name________________________________

ADDITION, SUBTRACTION,
MULTIPLICATION

Use the price list. Solve each problem.

1.

Total $____________

4.

$____________

$____________

$____________

Total $____________

2.

Pat bought 

Mike bought 

How much more
did Mike spend? $____________

5.

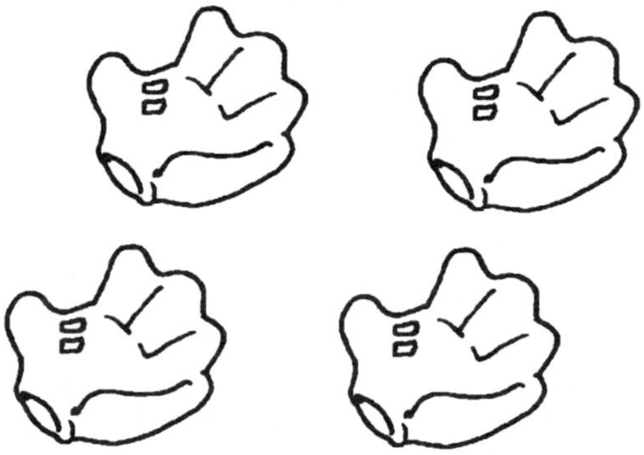

Total $____________

3.

$____________

$____________

$____________

$____________

Total $____________

6.

Bonnie bought 

Betty bought 

How much more
did Betty spend? $____________

Name________________________________

ADDITION, SUBTRACTION,
MULTIPLICATION

Use the price list. Solve each problem.

1. Lisa bought a watch, a book, and a pair of sunglasses. How much did she spend? $______________

2. Find the cost of 3 pairs of earrings and 3 bracelets. $______________

3. Eric had $7.48. He bought a kite and 2 coloring books. How much is left? $______________

4. Mrs. Fisher bought a cookie jar and a basket. How much change did she get from $15.00? $______________

5. How much more is a baseball glove than a basketball? $______________

6. Mr. Baker bought a backpack, a fishing pole, and jeans. How much did he spend? $______________

7. Find the cost of 5 wallets. $______________

Name________________________________

ADDITION, SUBTRACTION, MULTIPLICATION

Use the price list. Solve each problem.

---

1. Find the cost of 8 

. $______________

2. You have $10.49.

   You bought 2  . How much left? $______________

3. How much more for a  than a  ? $______________

4. Libby bought   and  .

   How much did she spend? $______________

5. Jake has $24.35. He wants to buy  .

   How much more does he need? $______________

6. Susan bought 2  and a  .

   How much did she spend? $______________

Name____________________________________

ADDITION - 3 items

Use the price list. Solve each problem.

1. Choose 3 items you would buy for your teacher.

____________________ $__________

____________________ $__________

____________________ $__________

Total $__________

2. Choose 3 items you would buy for a friend.

____________________ $__________

____________________ $__________

____________________ $__________

Total $__________

3. Choose 3 items you would buy for yourself.

____________________ $__________

____________________ $__________

____________________ $__________

Total $__________

4. Choose 3 things to wear.

____________________ $__________

____________________ $__________

____________________ $__________

Total $__________

5. Choose 3 sport items.

____________________ $__________

____________________ $__________

____________________ $__________

Total $__________

6. Choose 3 things to wear.

____________________ $__________

____________________ $__________

____________________ $__________

Total $__________

ANSWER KEY

Page 1 1. $1.50 2. $9.90 3. $11.00 4. $8.50 5. $3.00 6. $6.00 7. $7.50 8. $3.98 9. $3.77 10. $13.75 11. $4.50 12. $3.75 13. $9.79 14. $10.00 15. $2.29 16. $7.99 17. $3.50 18. $3.49 19. $6.44 20. $2.49 21. $14.35 22. $3.50.

Page 2 1. $6.05 2. $7.50 3. $8.44 4. $7.49 5. $.50 6. $9.44 7. $1.25 8. $4.95 9. $34.00 10. $10.98 11. $4.99 12. $11.00 13. $3.00 14. $8.49 15. $12.00 16. $9.25 17. $7.49 18. $3.29 19. $1.60 20. $8.77 21. $12.00 22. $10.49

Page 3 1. $7.50, $2.49, $9.99 2. $3.00, $4.95, $7.95 3. $8.44, $1.50, $9.94 4. $6.00, $10.00, $16.00 5. $11.00, $2.29. $13.29 6. $3.50, $3.29, $6.79 7. $7.50, $25.00, $32.50 8. $14.35, $12.00, $26.35

Page 4 1. $6.99 2. $2.85 3. $15.90 4. $10.49 5. $10.55 6. $22.00 7. $37.50 8. $2.79

Page 5 1. $7.79 2. $16.95 3. $18.99 4. $20.98 5. $12.79 6. $37.98 7. $12.90 8. $16.94 9. $10.99 10. $15.69

Page 6 1. $14.77 2. $10.99 3. $25.75 4. $12.99 5. $21.98 6. $25.50 7. $10.99 8. $14.99

Page 7 1. $8.77, $12.00, $20.77 2. $3.75, $9.44, $13.19 3. $10.98, $4.50, $15.48 4. $2.29, $3.29, $5.58 5. $10.49, $1.25, $11.74 6. $9.90, $3.49, $13.39 7. $8.49, $9.25, $17.74 8. $8.44, $7.49, $15.93

Page 8 1. $10.78 2. $15.00 3. $12.54 4. $12.00 5. $15.48 6. $15.49 7. $17.50 8. $4.09 9. $5.49 10. $21.79

Page 9 1. $8.44, $7.49, $15.93 2. $3.29, $2.29, $5.58 3. $4.50, $13.75, $18.25 4. $7.49, $4.95, $12.44 5. $10.98, $7.49, $18.47 6. $34.00, $6.05, $40.05 7. $9.90, $3.50, $13.40 8. $10.49, $2.29, $12.78

Page 10 1. $33.50 2. $20.40 3. $9.00 4. $15.88 5. $4.09 6. $12.45 7. $15.44 8. $18.39

Page 11 1. $7.50, $10.00, $1.25, $18.75 2. $3.49, $1.50, $12.00, $16.99 3. $11.00, $3.50, $2.49, $16.99 4. $12.00, $4.50, $3.29, $19.79 5. $11.00, $2.29, $.50, $13.79 6. $10.98, $6.00, $3.00, $19.98

Page 12 1. $16.55 2. $35.00 3. $48.50 4. $15.89 5. $8.75 6. $16.79 7. $11.29 8. $26.85

Page 13 1. $10.50 2. $37.89 3. $14.94 4. $18.79 5. $25.29 6. $22.00

Page 14 1. $8.50, $1.25, $10.00, $19.75 2. $3.98, $11.00, $25.00, $39.98 3. $11.00, $6.44, $.50, $17.94 4. $12.00, $3.29, $3.50, $18.79 5. $34.00, $10.98, $3.00, $47.98 6. $6.00, $1.50, $2.29, $9.79

Page 15 1. $7.50, $13.75, $11.00, $32.25 2. $8.77, $12.00, $11.00, $31.77 3. $4.50, $9.90, $6.05, $20.45 4. $14.35, $4.99, $10.00, $29.34 5. $9.44, $6.44, $7.49, $23.37 6. $4.95, $3.77, $12.00, $20.72

Page 16 1. $16.74 2. $16.55 3. $30.28 4. $19.48 5. $70.00 6. $4.39 7. $23.21 8. $23.29

Page 17 1. $19.98 2. $13.50 3. $20.98 4. $5.28 5. $16.92 6. $38.45

Page 18 1. $16.99 2. $31.30 3. $22.24 4. $11.50 5. $30.50 6. $19.76 7. $14.74 8. $19.29

Page 19 1. $2.79 2. $12.50 3. $21.24 4. $16.05 5. $22.00 6. $8.56 7. $24.37 8. $16.27

Page 20 1. $2.00 2. $5.00 3. $2.10 4. $3.34 5. $2.20 6. $.20 7. $1.23 8. $1.59

Page 21 1. $1.40 2. $6.39 3. $2.50 4. $2.75 5. $5.47 6. $8.29 7. $6.40 8. $2.39

Page 22 1. $13.00 2. $2.32 3. $1.10 4. $4.35 5. $.40 6. $4.00 7. $3.20 8. $3.11

Page 23 1. $1.75 2. $.35 3. $3.00 4. $1.00 5. $2.70 6. $7.20

Page 24 1. $6.41 2. $6.26 3. $1.75 4. $3.99 5. $4.67 6. $7.50

Page 25 1. $8.44, $6.00, $2.44 2. $34.00, $10.49, $23.51 3. $13.75, $8.77, $4.98 4. $14.35, $9.79, $4.56 5. $10.98, $25.00, $14.02 6. $9.25, $4.99, $4.26

Page 26 1. $.75 2. $2.69 3. $2.65 4. $.89 5. $.52 6. $3.85 7. $2.26 8. $.87

Page 27 1. $3.50 2. $2.56 3. $.80 4. $2.23 5. $1.29 6. $1.85

Page 28 1. $3.00 2. $6.26 3. $4.00 4. $1.25 5. $1.50 6. $3.00

Page 29 1. a) $6.10 b) $.90 2. a) $14.77 b) $.23 3. a) $11.00 b) $1.00 4. a) $25.75 b) $14.25

Page 30 1. $2.20 2. $5.00 3. $1.50 4. $3.35 5. $1.45 6. $6.05 7. $5.80 8. $2.22

Page 31 1. $10.50 2. $6.26 3. $2.06 4. $1.21 5. $.73 6. $1.55 7. $2.11 8. $.00

Page 32 1. $3.01 2. $2.00 3. $7.35 4. $.71 5. $4.50 6. $4.61 7. $.50 8. $.12

Page 33 1. $1.69 2. $.28 3. $1.50 4. $2.51 5. $1.00 6. $2.06

Page 34 1. $3.00 2. $1.50 3. $9.00 4. $18.00 5. $3.20 6. $36.00 7. $9.00 8. $22.00 9. $3.75 10. $24.00

Page 35 1. $16.88 2. $22.50 3. $24.00 4. $ 2.50 5. $14.00 6. $10.50 7. $48.00 8. $7.50 9. $68.00 10. $13.50

Page 36 1. $6.00 2. $3.00 3. $4.58 4. $60.00 5. $4.80 6. $27.50 7. 7.50 8. $42.50 9. $18.88 10. $30.00

Page 37 1. $75.00 2. $9.16 3. $3.50 4. $14.98 5. $19.90 6. $19.96 7. $22.50 8. $9.00

Page 38 1. $12.00 2. $14.94 3. $19.58 4. $14.97 5. $37.00 6. $31.47 7. $6.25 8. $27.50

Page 39 1. $4.50 2. $22.47 3. $10.98, $3.77, $3.50, $18.25 4. $3.98, $2.29, $2.49, $8.76 5. $3.44 6. $44.00

Page 40 1. $21.05 2. $.98 3. $18.00 4. $10.98 5. $23.51 6. $.70

Page 41 1. $18.27 2. $4.22 3. $9.77 4. $7.07 5. $3.12 6. $20.10

Page 42 1. $17.50 2. $5.58 3. $6.05, $9.79, $10.00, $6.00, $31.84 4. $8.49, $3.77, $3.00, $15.26 5. $100.00 6. $23.00

Page 43 1. $17.58 2. $18.00 3. $3.99 4. $1.61 5. $13.00 6. $28.75 7. $30.25

Page 44 1. $4.00 2. $3.49 3. $3.99 4. $19.72 5. $9.65 6. $18.33

Page 45 Answers will vary.

